AF482145

This book is
ultimately dedicated
to our baby girl
Amara Celestine J. Basa,
sons
Niel Michael J. Basa
and
Jasper Miguel J. Basa.

INTRODUCTION

Have fun with Numbers is an introduction to numbers from 1 to 10 by using images. Toddlers will learn to count in this book by using colorful images that will catch their attention.

This book will also help kids write for the first time by tracing the numbers.

Read the Numbers

0   1   2

3   4   5

6   7   8

9   10

1 - one

Trace the number 1

1 1 1 1 1 1 1 1 1

Count the rabbit

1 flower   1 leaf

1 tree   1 stone

1 doll   1 coin

2 - two

Trace the number 2

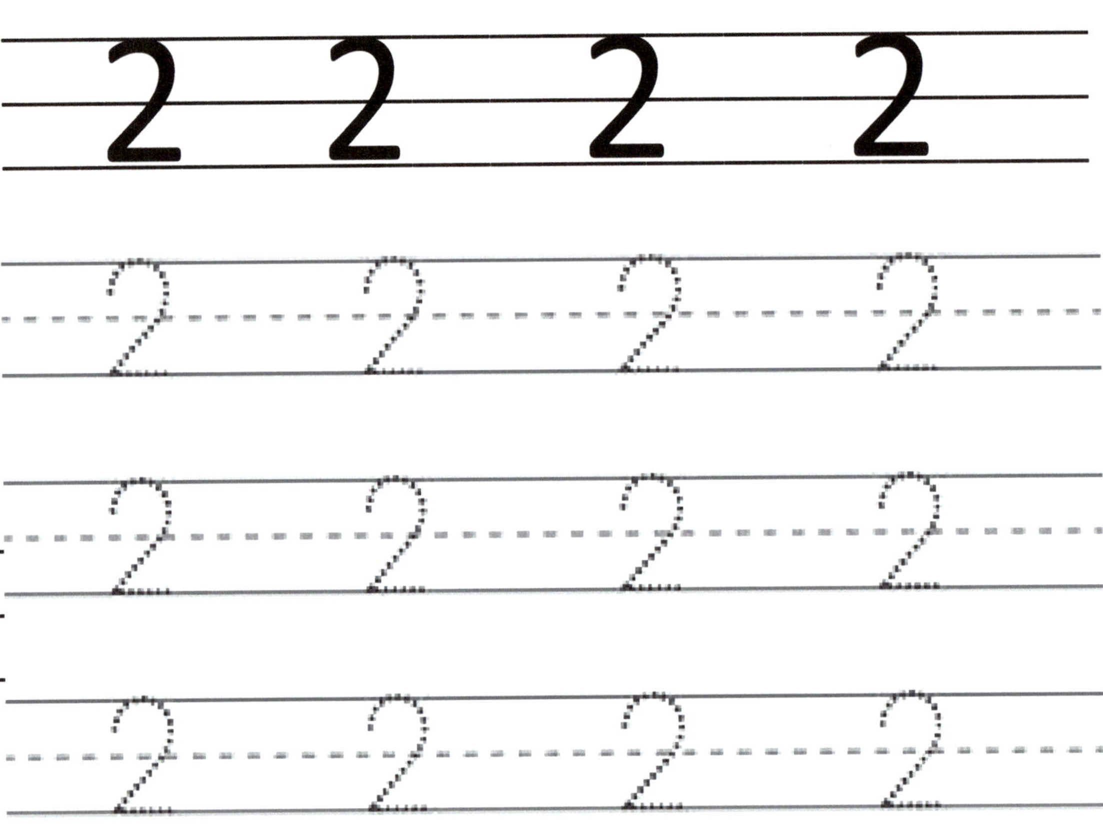

Count the zebras

Count the animals and trace the number 2

2 rabbits   **2 cats**

2 rats   **2 dogs**

2 cows   **2 snails**

3 - three

Trace the number 3

3　　3　　3　　3

Count the monkeys

Count the images and trace the number 3

3 ducks   3 hen

3 cats   3 birds

3 wolves   3 horses

4 - four

Trace the number 4

Count the ducks

4 camels

4 sheeps

4 hippos

5 - five

Trace the number 5

5   5   5   5   5

Count the Butterflies

5 ants

5 crickets

5 bugs

6 - six

Trace the number 6

6   6   6   6   6

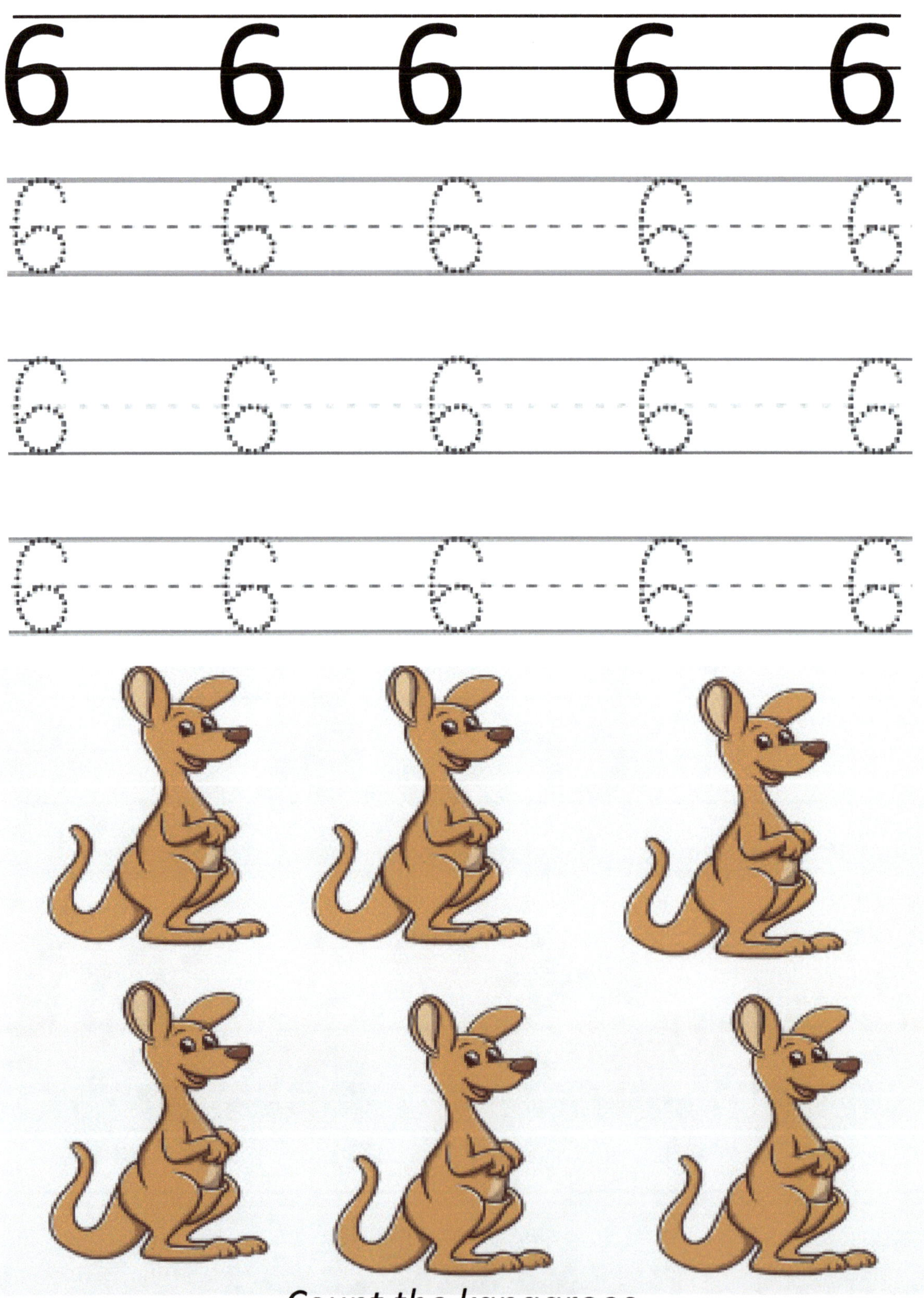

Count the kangaroos

6 foxes

6 bears

6 Lions

7 - seven

Trace the number 7

Count the pandas

7 giraffes

7 tigers

7 lizards

8 - eight

Trace number 8

Count the frogs

Count the animals images and trace number 8

8 pandas

8 dolphins

8 crocodiles

9 - nine

Trace number 9

9 9 9 9 9

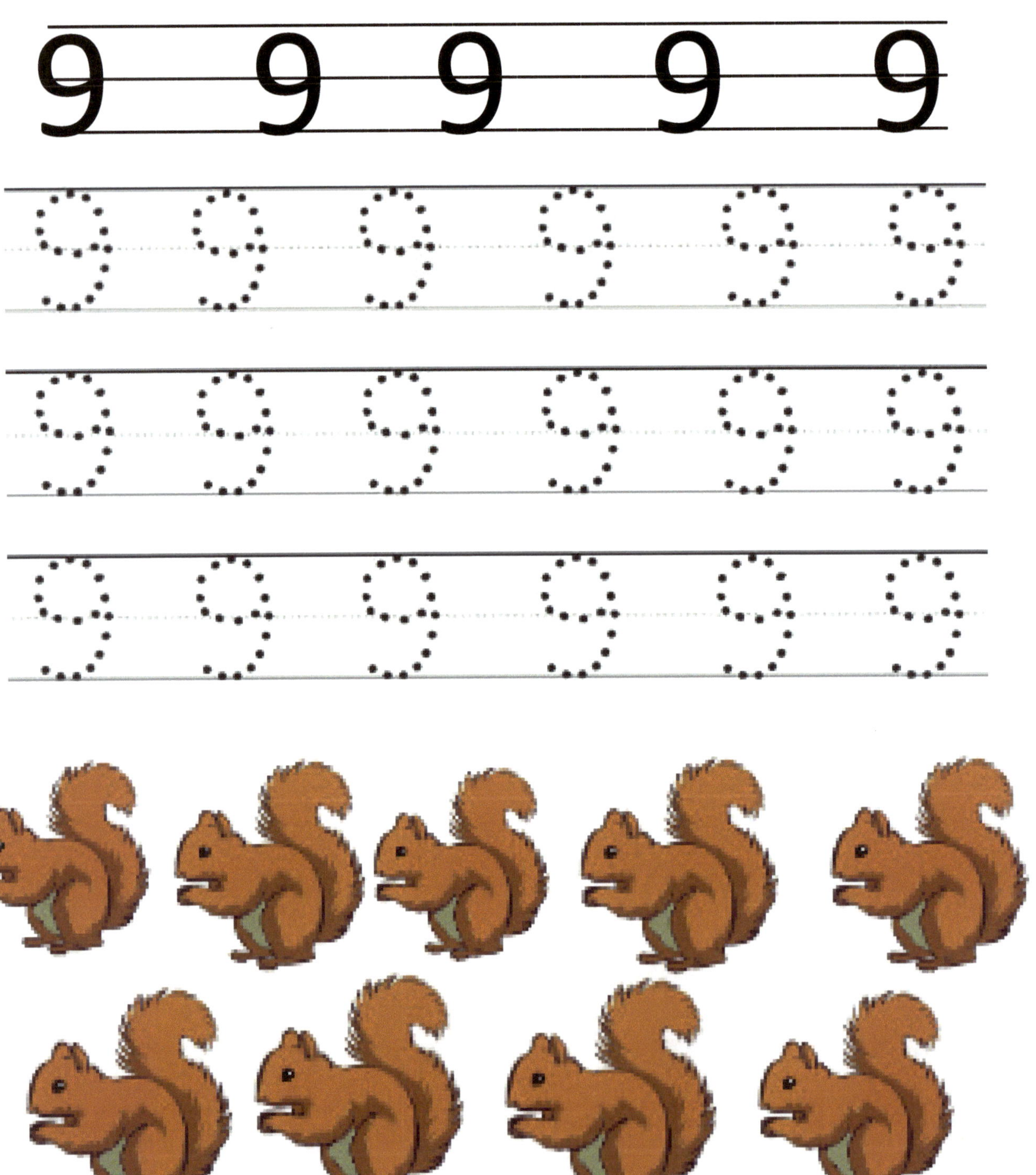

Count the squirrels

9 starfishes

9 octopuses

9 squids

10 - ten

Count the images and trace the number 9

10 10 10 10

10 10 10 10 10

10 10 10 10 10

10 10 10 10 10

10 10 10 10 10

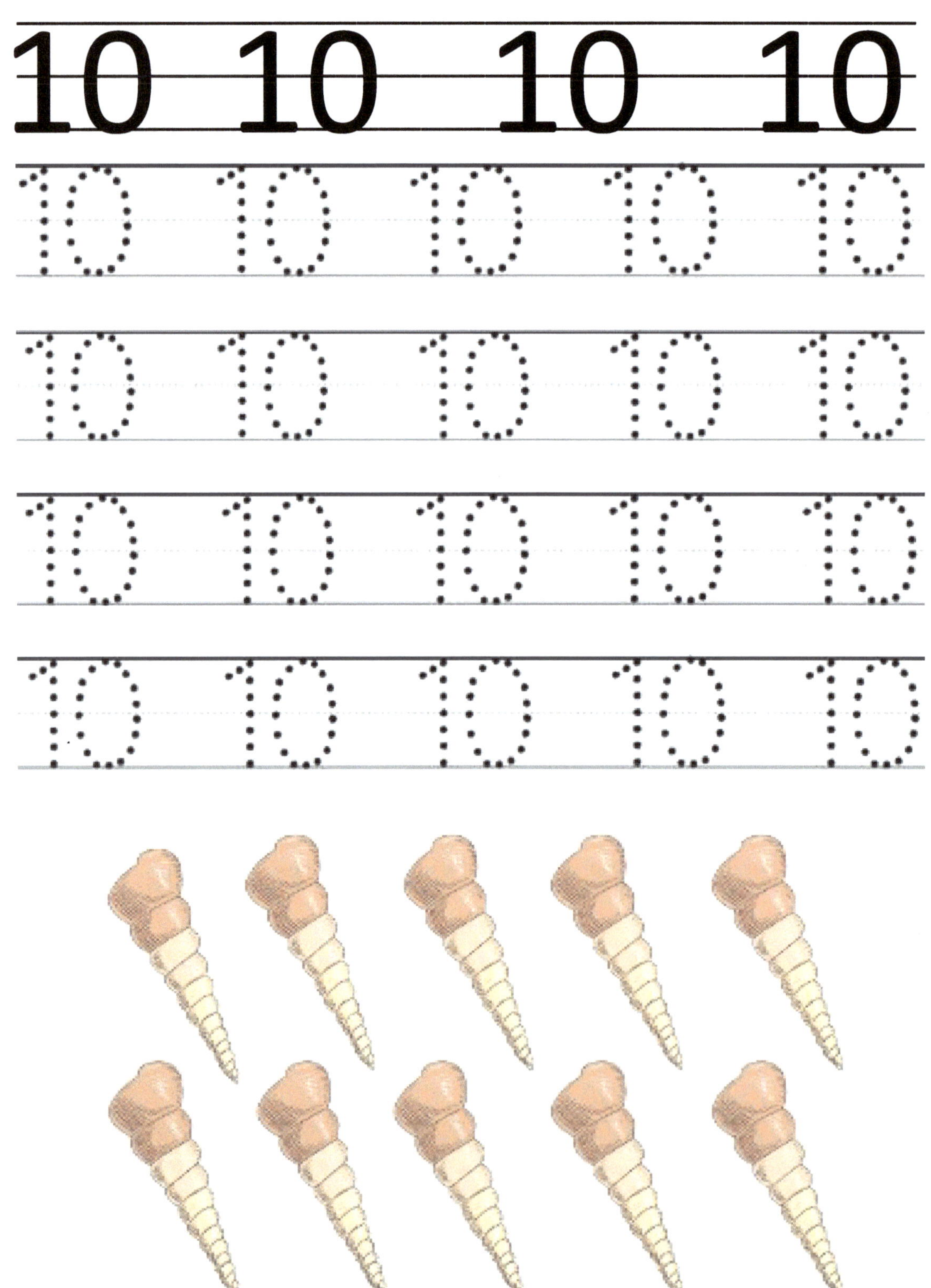

10 turtles

10 otters

10 walruses

Count the oranges and trace the numbers

Oranges	Trace
 🍊	1 1 1 1 1 1
 🍊🍊	2 2 2 2 2 2
 🍊🍊🍊	3 3 3 3 3 3
 🍊🍊🍊🍊	4 4 4 4 4 4
 🍊🍊🍊🍊🍊	5 5 5 5 5 5
 🍊🍊🍊🍊🍊🍊	6 6 6 6 6 6
 🍊🍊🍊🍊🍊🍊🍊	7 7 7 7 7 7
 🍊🍊🍊🍊🍊🍊🍊🍊	8 8 8 8 8 8
 🍊🍊🍊🍊🍊🍊🍊🍊🍊	9 9 9 9 9 9
 🍊🍊🍊🍊🍊🍊🍊🍊🍊🍊	10 10 10 10 10 10

Exercises

Color the correct number of apples.
Number 1 is done for you.

Color the correct number of apples.
Number 6 is done for you.

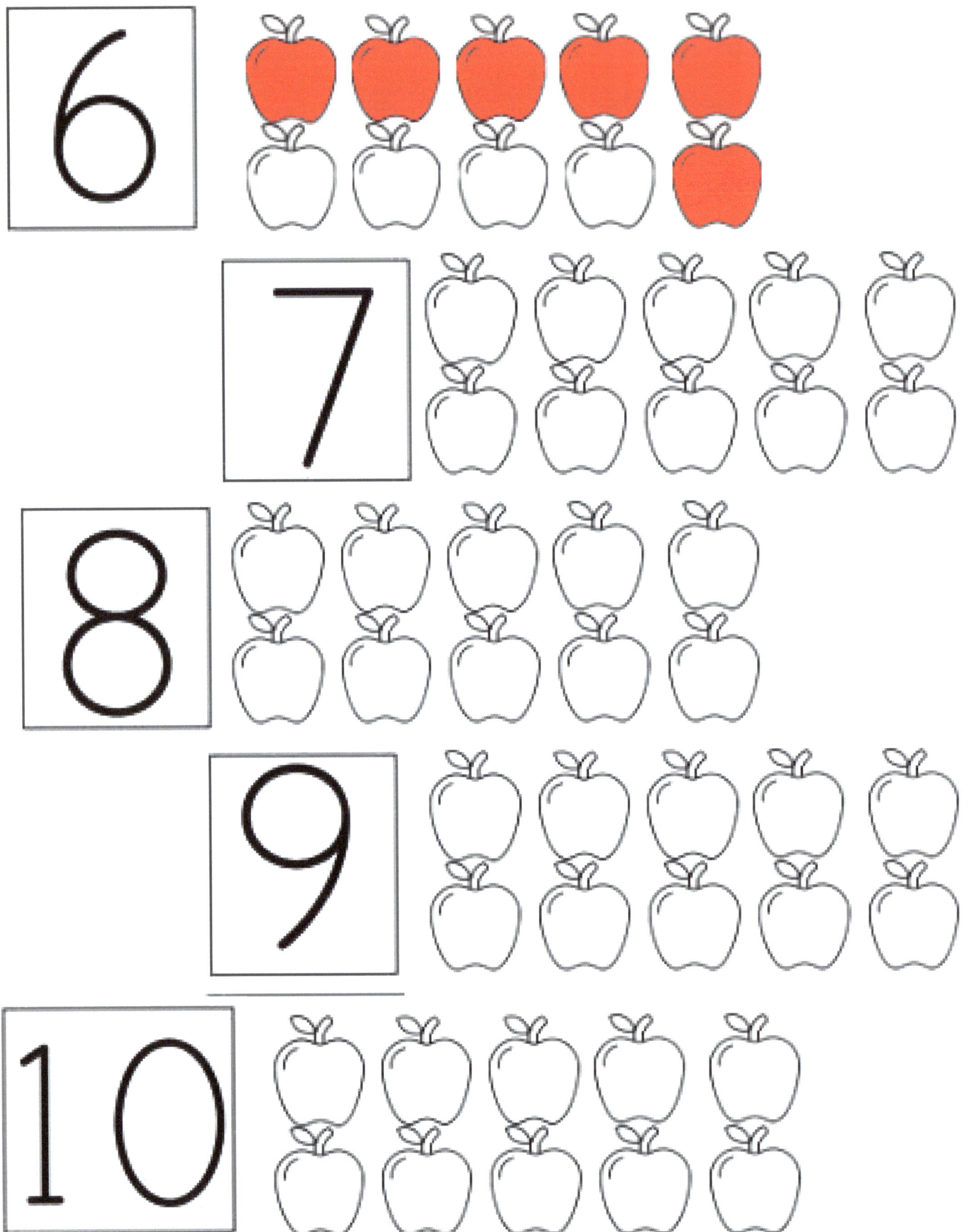

**COPYRIGHT © HAVE FUN WITH ALPHABETS
By ROY BASA**

All rights reserved. No part of this publication may be reproduced, distributed, or transmitted in any form or any means, including photocopying, recording, or other electronic or mechanical methods without the prior permission of the publisher and author, except in the case of brief quotations embodied in critical reviews and certain other commercial uses permitted by copyright law may be reproduced or used in any manner without the prior written permission of the copyright owner and publisher, except for the use or brief quotations.

ISBN
Hardbound-978-621-470-336-4
Softbound/Paperback-978-621-470-337-1
MOBI/KINDLE-978-621-470-338-8

**Published by:
Poetry Planet Book Publishing House
Rosario, Pozorrubio, Pangasinan, Philippines
Contact No.: 09554960044
Email: maritesritumalta@gmail.co**

Roy B. Basa
LPT, PhD, DHum, DMin, DSc, FPOd, FRIEdr

Roy Basa was born and raised in Murcia, Negros Occidental, Philippines by Raul and Lilia Basa together with his other 6 siblings. He graduated his elementary education from Lopez Jaena Elementary School.

He then went to La Consolacion College, Murcia for his secondary education where he graduated as Class Valedictorian. He got his Bachelor's in Education major in General Science, Master's in School Administration and Supervision, and Doctor of Philosophy major in Educational Management at the University of Negros Occidental – Recoletos where he graduated with Outstanding Dissertation and High Academic Distinction Awards. He then took another masterate, the Master in Natural Science at the University of St. La Salle, Bacolod under a scholarship grant, Project – Free Paglaum.

He was a high school, college, and graduate school science teacher for 18 years in the Philippines and 3 years as a high school science teacher in Arizona and New Mexico, USA, respectively.

He was awarded as one of the Most Outstanding Teachers of the Philippines in 2016 by the Metrobank Foundation, Philippines. Recently, he was also awarded by Asia – Pacific Luminare Awards as "Asia's Most Remarkable and Exceptional Science and CTE Educator of the Year 2022.

ROZEL JAENA BASA, MBA

Rozel Jaena Basa was born in Bacolod City and raised in Murcia, Negros Occidental, Philippines by Romeo and Razel Jaena together with her other 3 siblings. She graduated with her elementary education from Murcia Elementary School. She then went to La Consolacion College, Murcia for her secondary education. She got her Bachelor of Science in Information Management and Master's in Business Administration at the University of Negros Occidental – Recoletos

She was a former Manager at Golden Sun Finance Corporation, Bacolod City, Philippines for 16 years. Recently, she is one of the Educational Assistants for Pre – K to 2 at Shiwi Ts'ana Elementary School, Zuni, New Mexico, USA.

www.ingramcontent.com/pod-product-compliance
Lightning Source LLC
Chambersburg PA
CBHW042123110726
48006CB00003B/749